LITTLE BOOK OF
DIGGERS

Ellie Charleston

LITTLE BOOK OF
DIGGERS

First published in the UK in 2012

© G2 Entertainment Limited 2012

www.G2ent.co.uk

Printed and bound printed in the China

ISBN 978-1-907803-29-1

Contents

Preface

As I was writing this *Little Book of Diggers*, I found myself often being reminded of the fact that as children most of us are fascinated by diggers, dumpers and mega machines. Do you remember as a child being in a play park or at a play farm and having to wait your turn for the most popular toy in the sandpit – the digger? If you were fortunate enough you may have had your very own digger in the back garden and have been the envy of your friends. I am sure our interest in these machines does not wane as we get older; it is merely hidden by the stresses and strains of adult life.

Whether you are a child or adult, I hope this little book will reignite your interest in diggers and with the many names both past and present that have become synonymous with the word 'digger' – Bucyrus, Marion, JCB, Liebherr, Caterpillar, New Holland and Komatsu, to name just a few.

The Little Book of Diggers does not purport to be a scholarly work. It gives a light-hearted potted history of the machines and their makers, from the early days when they were known as steam shovels right through to the

advanced machines in production today. Finally, it gives a glimpse into what the future may hold for these machines.

I would like to thank Ray Hooley for his advice, in particular for imparting his knowledge of Ruston Hornsby engines.

Introduction

Man has been constructing buildings and other structures since ancient times. And we only have to think of the Pyramids of Egypt and Central and South America, Stonehenge and Avebury in England, to name just a few, to realise the staggering amount of manpower, with some assistance from beasts of burden, that would have been needed to build them. But it was not until the rise of the great civilisations of Greece and Rome that the first examples of mechanised construction and rudimentary machines began to be used.

Perhaps one of the best known historical mechanical shovels is depicted in a sketch made by Leonardo da Vinci in 1513. The sketch shows a dredger suitable for cleaning out a riverbed or canal bottom. The machine is placed on two boats and between the boats is mounted a revolving wheel equipped with four buckets. The buckets collect the mud and drop it on a raft that floats between the boats. The dredge moves

forward by way of a mooring cable which coils round the drum as the wheel turns. However, it was not until the industrial revolution in the 18th and 19th centuries, when true mechanical forms of construction equipment were considered and invented, that the face of the construction industry changed for ever.

The idea of using boiling water to generate power had been considered thousands of years before but it had not been found to be practical or efficient. But with improved developments and increases in power, steam began to replace muscle power as the primary source of power. All manner of steam-powered machinery was used in industrial settings and steam-powered vehicles became a practical proposition.

The construction industry was quite late in catching on with this new technology; steam cars made an appearance long before steam excavators. But when they were accepted, steam shovels, or steam-driven mechanical excavators, became some of the most important contributions in the development and evolution of earthmoving machines.

Steam Shovels

The earliest type of mechanical digger was the steam shovel. In Great Britain during the industrial revolution, steam shovels really came into their own with the construction of the canal and railway network, and a similar thing happened in America with the building of new roads, factories, docks, ports and buildings.

But it was not just in construction that diggers proved to be of use; mining and quarrying also benefited from their introduction and as a result they were used throughout the world, wherever mining operations took place.

Steam shovels usually – though not always – had a three-man crew: an engineer, a fireman and a crane man. The engineer was often also called the driver as he drove the machine along the track and also raised and lowered the bucket. The fireman kept the boiler topped up with coal and water and the crane man had the job of sitting to the side of the boom and tipping the bucket.

The major components in a steam shovel are:
• bucket
• water tank
• winch
• coal bunker
• steam engines
• boom
• wheels, caterpillar tracks or railway tracks
• dipper stick

The steam shovel was the forerunner of all powered, single-bucket, 'dry land' excavators and was the first machine to replace the hand shoveller on construction sites.

William Smith Otis, an American civil engineer, received the first patent for a steam shovel. In 1839, patent No 1,089 was granted to him for his design for a machine that excavated and removed earth. His patent drawing shows a crane mounted on a railway carriage. Earth could be lifted by the bucket, raised by the crane and turned to be dumped. The patent described how a steam engine of

a type then in ordinary use was installed with a power control mechanism for the crane, and a system of pulleys to move its arms and the bucket.

The machine could move about 380 cubic metres of earth a day with its 1.1 cubic metre capacity shovel. It was first used on the Western Railroad before being moved to Canada, where it was used on a pontoon to excavate canals. Records suggest that this first digger was finally broken up in 1905.

William Otis was a cousin of Elisha Otis, who was famous for his invention of the lift. From an early age, William showed a great deal of interest in earthworks and mechanics, and the idea for his history-changing concept came when he was working with the company Carmichael and Fairbanks on a railroad project that involved a great deal of time-consuming and back-breaking earthmoving activity. Otis was convinced there had to be a more efficient way to remove all that earth, and he devised an apparatus that would carry out the same actions as a person with a shovel.

By the age of 22, Otis had constructed a prototype. Realising that on his own he would be unable to get a production

model into existence, he moved to Philadelphia. There he obtained the support of Joseph Harrison, who operated the railway locomotive company of Garrett and Eastwick, a company that was involved in the expansion of the railroad system of the American Midwest. He joined the company and, with Harrison's assistance, constructed a pre-production model in 1836. The machine consisted of a vertical steam engine fastened to a boiler and a central mast, on to which was attached the boom, which swung at 180 degrees. By today's standards, this model would be judged a crude and clumsy device but at the time it was considered revolutionary.

On 15 June 1836, Otis received a patent for this invention, but unfortunately the engineering specifications were destroyed during a fire. It took another three years before the patent behind No 1,089 became officially valid. This patent was called 'the crane-dredge for excavation and earth removals'.

In 1839, Otis contracted typhoid while working with his shovel for the Western Railroad of Massachusetts, and he died aged just 26. It was in that same year that

STEAM SHOVELS

he received his patent, before he could see the success his invention would achieve. His shovel brand was taken over by Oliver Chapman and the machines became known as Otis–Chapman shovels.

Because Otis had patented his steam shovel, other equipment manufacturing companies could not benefit from the device until the patent expired in the 1870s. Otis's second and third shovels were dispatched to Russia and his fourth machine was used in the construction of the Eastern Counties railway in 1840.

The first machines were known as 'partial-swing'; the swing was partial because the dipper arm could not rotate through 360 degrees. They were built on railway chassis on which the boiler and movement engines were mounted. The shovel arm and driving engines were mounted at one end of the chassis. Bogies with flanged wheels were fitted, and power was taken to the wheels by a chain drive to the axles.

Where the shovel was expected to work, workers laid temporary rail tracks, to be repositioned as and when required. Later machines were supplied with caterpillar tracks so the tedious job of laying temporary tracks was

Shovel At Conveyor
5th Ave. & Battery St.

7483
5-11-29

00 Ton Steam Shovel in Operation.   Michigan Limestone and Chemical Co.   CALCITE, M

PUBL. BY
HENRY UHL

avoided. By the 1870s, the hoists were being made out of steel, allowing for easier rigging to the winches.

As the railway network expanded both in the UK and across the pond, the demand for steam shovels increased and as a result they became commonplace. Companies such as Bucyrus and the Marion Steam Shovel Co were formed and moved into the manufacture of steam shovels.

The Marion Steam Shovel Company was founded by Henry Barnhart, George W King and Edward Huber in 1884, in Marion, Ohio. Barnhart had plans for a bucket that was stronger than those already in production, and he persuaded his partner Huber to provide financial backing for his new design, with its solid iron rods that supported the boom of the shovel. In 1883, Barnhart and Huber patented Barnhart's changes under United States Patent No 285,100.

Marion's shovels gained a reputation for efficiency and in July 1908 one of their machines set a record by moving 53,000 cubic yards of earth in 25 eight-hour days. Then, three years later, another Marion machine broke the world record for lifting the greatest amount, eight tons, in a single bucket. By 1911, 90 per cent of all large bucket steam shovels were produced by the Marion Steam Shovel Company. From 1900 to the 1960s, Marion-built shovels were considered to be one of the best-known trade names in earthmoving equipment.

The city of Marion was also the headquarters of the Osgood Steam Shovel, Fairbanks Steam Shovel and General Excavating Corporation companies. Their main rival, the Bucyrus Steam Shovel Company, was founded just 15 miles away in Bucyrus, Ohio.

Bucyrus was founded in 1880 and 13 years later moved to South Milwaukee, Wisconsin. Soon after, the company directors decided to specialise in building excavating machinery, and the decision paid off. The company won the contract to build 77 of the 102 steam shovels used in the construction of the Panama Canal between 1904 and 1914, after Count Ferdinand de Lesseps' company, initially engaged in its construction, ran into financial difficulties. The remainder of the contract was given to rival company Marion. Machinery was needed to cut a 51-mile long stretch from the canal's entrance in the Atlantic through to its

LEFT
A 100 ton steam shovel, circa 1919

STEAM SHOVELS

RIGHT
A Ruston-
Bucyrus shovel
from the
1930s. Credit
BulldozerD11

terminus in the bay of Panama.

The departing French company had left a great deal of machinery behind, but it was replaced by far superior machines designed for a larger scale of work. Despite this improvement, the work was still very labour intensive as five men were needed to operate these particular steam shovels – a crane man, a main operator, a fireman and two wheelmen. The sixth person in the system was the mechanic. The excavation teams from Bucyrus and Marion would race to see who could excavate the most material. When the Panama Canal was completed in 1914, the steam shovels were sent to Montana, Ohio, Alaska, Spain and Costa Rica to work in mining fields.

Following the success of its involvement in the Panama Canal project, Bucyrus became the market leader. Various acquisitions took place over the years, with one of the first being the purchase of rival business the Vulcan Steam Shovel Company. In 1927, Bucyrus merged with the Erie Steam Shovel company, a well-established manufacturer, to form the Bucyrus Erie Company.

Three years later, Bucyrus joined forces with the English firm of Ruston & Hornsby of Lincoln, who were well established in the world of steam shovels, and the firm of Ruston Bucyrus was formed. Ruston & Hornsby's entire excavator manufacturing operation was transferred to Bucyrus, but the merger enabled Bucyrus to sell into European markets; something that had previously been difficult for them. In 1997 Bucyrus acquired the Marion Power Shovel Company, ending an intense competitive rivalry between the two companies that had lasted over 100 years.

Machines were also being built in Great Britain, and one manufacturer who had foreseen the importance of these early steam 'navvies' was Joseph Ruston. Born at Chatteris, Cambridgeshire in 1835, he entered into a partnership with the firm of Messrs Burton & Proctor of Lincoln, builders and repairers of all kinds of agricultural machines and implements, in 1856. They needed a salesman to expand their business and the ambitious Ruston was just the person.

In the 1880s he travelled to Russia to negotiate a deal for steam engines and pumps to drain eight million acres of the Pripet Marshes. A few years later,

Ruston persuaded a group of Lancashire businessmen that a ship canal from Liverpool to Manchester would be an economic viability. He pointed out that previously, Britain's canals and railways had been largely built by manual labour, known as navvies, but if Ruston-built mechanical excavators, or 'steam navvies', were used, the time taken to cut a channel would be greatly reduced. In all, 71 of these machines were supplied to the Manchester Ship Canal Company when construction commenced in 1887.

The Ruston steam navvy was far

superior to other machines of the time as it used steel beams instead of timber ones, and so was able to dig out a wide range of material, be it sand, clay, gravel or chalk, far more efficiently. The steam navvy scooped up material with its large bucket, then swung sideways to empty the contents into a waiting wagon. Despite being of English origin, the Ruston navvies became known by the manual labourers as 'American Devils', apparently as the result of an earlier incident on a different project, where workers in conflict with the contractor were replaced by an American digging machine.

Priestman Brothers, another early manufacturer of construction equipment, was founded in 1870 in Kingston upon Hull by the Quakers William and Samuel Priestman. It is said that in the mid-1870s the brothers were asked to build equipment to look for lost gold off the coast of Spain. No gold was ever found, but the Priestman equipment was used to dredge harbours and docks, and was used soon after at Hull docks with such success that it was soon in demand all over the world.

Priestman's early machines were very rudimentary, with the driver sitting next to the engine. After World War I, they produced a machine for digging drainage ditches in fields: their first real foray into the world of diggers. Priestman Brothers then produced a number of diggers named after animals – Lion, Tiger and Panther – and these names became synonymous with the company.

Over the years there have been financial difficulties, takeovers and amalgamations, and today the company is owned by an American compressor company and no longer manufactures diggers.

All these early excavators were shovels of the mast or tower type, but a new type was on the way in the 1890s. Captain Richard P Thew of South Lorain, Ohio was the captain of a freighter on Lake Erie – named the William P Thew after his father – which transported iron ore to the steel mills of Cleveland. Thew was paid by commission; the more ore he loaded, the more he got paid. Realising that the system in use for moving the iron ore from his freighter to the docks was inefficient, he decided to invent a new kind of shovel.

Thew produced a fully revolving

shovel in which the machinery, including the boom, was mounted on a revolving frame and the whole entity had a 360 degree swing. Being able to swing in a complete circle enabled it to dig and deliver materials from any direction.

Thew's machine proved a great success and several other steel companies in the area decided they wanted one, so Thew, together with businessman F A Smythe, set up the Thew Automatic Shovel Company in 1899 to manufacture machines. Thew's machines continued to be a tremendous success and the company was soon building them at the rate of one a month. The Thew Automatic Shovel Company continued to develop its machinery, and in 1924 it produced a whole new line of power shovels under the trade name of Lorain. This is the name by which Thew's products have been known ever since.

Back in Great Britain, the first machine to revolve 360 degrees was built by Whitaker and Sons of Horsforth, Leeds in 1874. It was described as a steam navvy, although it was really a steam crane navvy: its excavating attachments could be removed to allow the machine to be used as a crane. It became affectionately known as the Jubilee Crane Navvy in honour of Queen Victoria's jubilee. Three years later, Whitaker and Sons followed up with the Wilson crane navvy, another machine built on a travelling crane.

Excavators started to be manufactured with wheels, thus eliminating the need for rail tracks to be employed whenever the need arose for a machine to be moved. The first wheeled excavators had small wheels and were mainly used on the railways. One of the earliest such examples was Little Giant, produced in the 1880s by the Vulcan Iron Works of Toledo, Ohio.

Further developments were taking place, with tracked vehicles proving to be far superior on rough terrain. These early tracked vehicles were steered from the ground, coupling together two shafts. Manufacturers were also increasingly starting to mount diggers on trucks, so that machines could be moved from one job to the next without having to be dismantled. Needless to say, this saved a lot of time, effort and man hours.

After Steam

By the 1930s, the rising cost of coal and the development of simpler and cheaper diesel-powered engines had caused steam shovels to fall out of favour. Diesel-operated excavators were far more fuel efficient, produced more horsepower and required fewer men to operate and maintain them.

To begin with, these engines were pretty slow, running at around 350 rpm. Working in dirty, dusty conditions and with exposed valves on their engines, they had a tendency to break down frequently. But further progress in, and development of, diesel engines continued throughout the period leading up to World War II and beyond. These early diesel machines became the forerunners of those still in use today.

After the war, far more refined machines began to be developed rapidly, with many more companies entering the construction equipment market; many names that are still with us today.

For the machines' drivers, though, there was very little change from the days of steam. With no enclosed cab, the driver had to work in the open air from a pedestal seat using mechanical controls to raise and lower the boom and manoeuvre the bucket. He would have been subject to loud noise and exposure to all sorts of weather conditions. This was, of course, in the days before there was as much concern about health and safety at work as we see today.

Initially, there were no gauges in the cab for the driver to check, but when they were first introduced they were all mechanical and were limited to oil pressure and temperature gauges. Sometimes, there would be an ammeter to measure the voltage.

Electric machines had been in existence since the late 1800s, and they were improved and developed upon as time went by, with various different systems of electrical control being used. One of the most interesting and enduring is the Ward Leonard motor control system. Though it was introduced in 1891, it is still used

nowadays; with refinements, obviously.

The system involves a motor, usually AC current turning at a constant speed and a generator. The generator, which is usually the DC motor, is responsible for adjusting the speed of the equipment, which it does by altering the voltage with the help of a rheostat. Both Bucyrus and Marion used the Ward Leonard system and eventually it became the general standard configuration for all electric shovels.

In the mining industry, cable shovels and backhoes became

popular, but by the 1950s the first fully hydraulic crawler-mounted excavator was introduced, and it really signalled the end of cable shovels and backhoes. In 1954, the first of these new hydraulic excavators to reach full production was the Demag B-504. Demag stands for Deutsche Maschinenfabrik AG – a German manufacturer of heavy equipment.

In 1971, three years after they opened their hydraulic plant in Verberie, the French company Poclain introduced the first large fully hydraulic mining excavator, named the EC1000. The first machine to break the 500 ton operating weight barrier was the RH300 built by Orenstein and Koppel of Dortmund.

Orenstein and Koppel, or O&K, was founded in Berlin in 1876 by Benno Orenstein and Arthur Koppel. The company originally built railway equipment, but by the start of the 20th century they had moved into building excavators and are now one of the largest mining shovel manufacturers in the world, holding about a 30 per cent share of the worldwide market.

In 1986, the 600 ton barrier was broken by the H485 made by another German firm, Mannesmann Demag, but it wasn't long before O&K were back, fabricating an even bigger machine. The RH400 was the world's largest, with track pads each two metres wide. The prototype was unveiled at O&K's factory in Dortmund in 1997.

After all the necessary factory tests were undertaken, the RH400 was dismantled and shipped across the Atlantic to its new home at the Syncrude North Mine project in Canada, where it was reassembled. It took ten days to put together again before the machine was ready for use. Since then, four more RH400s have been built and all have been fitted with a standard small kitchen with basic appliances, heating and air-conditioning.

Many of the early leviathans had unhappy endings and once they started to develop faults, they were deemed too expensive to repair and were cut up or scrapped. However, a few are on display in museums or in private collections.

Today, engineers are developing hybrid and bio-fuel powered excavators with numerous amenities for the comfort of the operator. But more about that in the final chapter.

LEFT
A backhoe digger during Operation Desert Storm 1991

Digger Types

For every kind of task or job there is a digger or excavator, and the machine that suits one application will not necessarily suit another. A digger suitable for stump removal or excavating a patio would be totally inappropriate for the tasks down a mine or in a quarry.

If the digger has to work in a wet environment, it is called a floating digger; in a tunnel it is called a boring machine. Some diggers are wheeled and some are tracked. Heavy machines will always be tracked as this enables a better distribution of the weight on the soil.

Cable Excavators

These machines developed out of the steam shovel but nowadays they are more or less obsolete. Although they are very much like other excavators, what marks them out as different is that the cables and wire ropes lower and raise the excavator arm and thus the bucket.

Many of the early Bucyrus and Marion machines were cable excavators.

The Backhoe Loader

The backhoe excavator is one of the most popular digging machines you will see on a building site nowadays.

Companies had been modifying farm tractors to carry a loader arrangement on the front and a backhoe at the back for many years, but it wasn't until 1957 that a wheeled tractor designed specifically for a backhoe was manufactured. Despite being based on a tractor, the backhoe loader is never referred to as a tractor.

The world's first fully integrated tractor loader/ backhoe was made by Case and called simply Model 320. This early model was compact and easy to manoeuvre as it had rubber tyres, and it was much in demand to carry out small jobs that had previously been done by manual labour. Although it had no cab, it was far more advanced than traditional tractors, even down to its padded seat with back support; tractors then had simple steel seats.

As the years passed, different

companies adapted and improved the backhoe exclusively for the construction industry, and diesel power, improved hydraulic linkages and four-wheel drive were introduced.

The backhoe works with a wide shovel at the front for scooping up material from above and a small metal bucket or backhoe at the back for digging out a trench, for example. It digs by drawing the earth backwards rather than lifting it forward, as happens when one digs with a spade. As the loader bucket moves up and down, the rams move in and out of their metal casing, which makes the excavator arm move.

The bucket will be fitted with metal teeth so that it can cut through the earth more easily. Sometimes the bucket will be split so that it can act as a grab.

The section of the arm closest to the vehicle is called the boom and the driver can make it longer or shorter to suit the job. The boom is attached to the vehicle

through the kingpost, and this allows the arm to move not just up and down but also sideways in an arc. The seat rotates to face the rear when the backhoe is in use.

Nowadays, these diggers are powered by hydraulics. Since the days of the basic steam-driven diggers, excavator design has moved on dramatically and there are now a number of different styles, including mono booms, which can only move up and down; ones with a knuckle boom, which move up and down as well as left to right; and hinged booms, which allow the boom to pivot in a semicircle.

The Backhoe

A backhoe consists of a hydraulically powered articulated arm with an excavating bucket attached at the end. The digging bucket is typically smaller than front-mounted loading buckets, but, due to the manoeuvrability provided by the arm, it is much more versatile.

The Excavator

While backhoe loaders are ideal for use on small projects and in confined spaces such as those found in urban engineering projects, larger tasks require the use of an excavator.

An excavator consists of boom, bucket and pivoting cab – often referred to as the 'house' – mounted on a tracked undercarriage. Extending from the cab is an articulated arm similar to a backhoe, although much larger in size.

All the functions of the excavator are now achieved through the use of hydraulic fluid.

Power Shovel

The power shovel, or stripping shovel, is a piece of machinery used for digging in the mining industry. It is basically the modern version of the steam shovel, although it is now usually powered by electric motors or on-board diesel engines.

The shovel consists of a hinged bucket, hung from a boom, that is used for digging. Nowadays there are very few manufacturers who make this monster of a machine, so their use has declined. At one time they were some of the heaviest mobile machines in the world, and the largest one of all was called Captain Shovel.

This giant was in fact the heaviest mining shoveller ever made. The bucket alone weighed 165 tons when empty, and when it was full it could hold 270 tons. Captain Shovel was six metres wide, more than five metres high and over eight metres deep. It had 36 electric motors, each one generating between 200 and 400 horsepower, and eight of them were required to lift the bucket. Eight crawlers were used to move the machine and each crawler was 15 metres long and over three metres wide. Sadly, Captain Shovel was destroyed by fire in 1991 and was sold for scrap.

Drag Line

Excavator work in a quarry requires an entirely different kind of digging machine to the one used on an urban building site. Generally, dragline excavators are used in quarries and mines as they can dig up much more than other kinds of digger.

DIGGER TYPES

These machines have a bucket, large enough to hold a family car, hung from the end of the boom by wires. These wires, called draglines, drag the bucket filled with extracted material towards the excavator – hence the term dragline. The hoist rope controls vertical movements while horizontal movements are controlled by the dragline.

Usually the bucket is lowered on to the material waiting to be excavated, and then pulled across by the dragline

before shovelling up the material and then being lifted by the hoist rope and taken to wherever the material is to be dumped. The drag rope then tips the bucket so that it empties. The basic principle of working these machines is very simple but they do require a high level of skill to operate.

The first dragline was built in 1904 by Page and Schnabe for use on the Chicago Drainage Canal. Draglines were limited for the first few years

ABOVE
Bucket wheel
excavator,
Germany.
Credit Raimond
Spekking

of their existence, until the invention of a walking mechanism that allowed the draglines to have mobility. Many different methods of mobility were tried in the early years, most without much success, but in 1913, the Martinson Tractor made its first appearance.

In that year, Oscar Martinson, of the Monighan Machine Company of Chicago, patented the first walking mechanism for a dragline, which became known as the Martinson Tractor Drive. This early, crude innovation consisted of two movable shoes on each side of the frame that enabled the dragline to become mobile. More than 200 of these

machines were installed between 1913 and 1926, when Martinson successfully patented a self-propelling method. The patent was purchased almost immediately by Bucyrus-Erie.

Many methods of getting these draglines to walk were tried by different manufacturers, but the first seriously successful American version was made by Page in the 1930s. In Great Britain, Ransomes and Rapier started to design and manufacture walking draglines. These machines were designed and constructed at the Ransomes and Rapier's Waterside Works in Ipswich, Suffolk but assembled at the sites

DIGGER TYPES

where they would work – usually
open cast mines in the Rutland and
Northamptonshire areas.

One particular dragline excavator
named Sundew, built by Ransomes &
Rapier and named after the winning
horse of the 1957 Grand National,
became something of a legend. When
it was built, it was the largest walking
dragline in the world, weighing 1675
tons and with a reach of 86 metres and
a bucket capacity of 27 tons. When iron
ore mining ceased in Rutland, it was
decided to move the machine to Corby.

Over a period of nine weeks in 1974,
this behemoth of a machine was 'walked'
an incredible 13 miles to the site north
of Corby. During the 'walk', the dragline
crossed three water mains, ten roads, a
railway line, two gas mains and a river. It
was all filmed for the children's television
programme *Blue Peter*, and Sundew
became very much part of the local
folklore after this extraordinary feat.

In America, the largest working
dragline is the Bucyrus-Erie 2570-WS
– and it, too, has made an incredible
journey. This machine weighs more than
150 Boeing 737 planes and is 220 feet
high with 'shoes' that measure 72 feet

in length by 14 feet wide. In August 2010, the dragline began an 18-mile 'walk' from one coal mine in Indiana, in the eastern United States, to the largest surface mine, Bear Run.

The machine travelled at less than one tenth of a mile per hour and the 'walk' did not stop at all for over a month as the machine crossed the Indiana countryside. At each roadway crossing, the support crew had to lay plastic, straw, clay and shale over the tar to protect the road surface. Once it was at Bear Run Mine, 2570–WS was put into operation straight away.

Bucket Wheel Excavator

These machines are used in the

mining industry and although they are often employed in open pits and in surface or strip mining, some are also used underground.

Although they are simple to operate, the larger machines, being up to 200 metres long and 100 metres high may require up to five men to work them. Some are so large that they have rest-room facilities for the operators' convenience.

A bucket wheel excavator consists of a large rotating wheel with numerous buckets placed around the rim that are used to scrape up the material as the wheel turns. These buckets have sharp edges so that they bite into the ground. Once the material has been scooped up by the buckets, the load is then tipped

out on to a moving track or conveyor belt inside the machine. A typical bucket wheel excavator can extract 40,000 bucket-loads, or an area the size of a football field but up to 30 metres deep in a single day.

These enormous machines move very slowly on huge crawler tracks and they operate best on soft soil without large boulders. They are much employed in the Ruhr mining industry of Germany.

The current record for the largest bucket wheel excavator is held by the RB293 manufactured by MAN Takraf. But the Bagger 288 built by the German company Krupp has an extraordinary history. When Bagger, a monster of a machine, was built in 1978, it became the largest tracked vehicle in the world. It is about 100 metres high and 250 metres long and took five years to design and manufacture.

Bagger 288 ('bagger' is the German word for digger or excavator) spent much of its working life at a strip mine in Hambach, southern Germany. When that mine neared the end of its life, it was decided to move the machine to a mine just over 22 kilometres away. It was deemed cheaper and easier to drive the machine there rather than dismantle it and move it piece by piece.

With a top speed of less than one kilometre an hour, the journey took three weeks to complete as the machine travelled over roads, traversed rivers and crossed railway lines. A team of 70 workers was required to keep the operation running smoothly and to ensure that as little damage as possible was made to the terrain the machine passed over.

Bucket chain excavators are very similar to bucket wheel excavators. Instead of the buckets being placed around the large wheel, they are attached to a shiftable bench conveyor driven by a powerful electric motor.

The largest bucket chain excavator ever built is the Tenova TAKRAF ERs 3750, which can cut to a depth of between 30 and 35 metres. This machine is still in production and, just like the wheel extractors, it is used in lignite mines.

Tunnel Diggers/Borers

Human beings have been digging tunnels for hundreds of years, sometimes in a conflict situation, sometimes for water systems, sometimes for

DIGGER TYPES

RIGHT Part of the cutter head from the front of the TBM used under the River Thames. Credit oyxman

underground transport. And for a long time the tunnelling was done by a man with a pick and shovel.

Tunnel Boring Machines, or TBMs, were developed using the principle of how the earthworm burrows into the ground, and they have been developed to work in all manner of geological conditions. Some of the biggest digging machines in the world are TBMs.

In England, the excavation for the London tube network was started using manual labour but in the 1890s, the City and South London route was the first to be bored out by machine. Builders dug through the earth with spades at the front as the machine moved forward.

Since then, tunnel diggers have been used in all manner of projects. One of the more interesting ones was the most recent and the only successful attempt at the Channel Tunnel between England and France, which is over 50km long. Eleven TBMs were needed on the project, and each one used during the construction was about 25 metres long and was capable of digging out the chalk at a rate of about five metres an hour. The largest amount of chalk dug out by a TBM in a single week was 426 metres.

These machines would dig out the chalk, collect the spoil and then transport it via conveyor belts to its final destination, which in England was Samphire Ho near Dover in Kent, which is now a nature reserve.

At the time of writing, an enormous tunnel is being built in Switzerland. When it's completed, at 57 kilometres long, it will replace Japan's 54 kilometre Seikan Tunnel as the world's longest vehicular tunnel and will cut the travelling time from Zurich to Milan by an hour.

The shell of the New Alpine Transversal (NEAT), as it is called, was finally cut through in October 2010 by a massive digging machine called Sissi, but this tunnel through the Alps will not be fully completed or operational until 2017, when high speed trains will link northern and southern Europe.

When the tunnel is completed, four Herrenknecht TBMs will have excavated more than 152 kilometres of shafts. On this particular project, Gripper TBMs are being used as they are technically suitable in hard rock situations.

TBMs are designed with a cutterhead at the front; the cutterhead spins around and cuts through the earth or rock

with sharp blades called disc cutters. In hard rock situations, the disc cutters break the rock into pieces around 15cm in diameter. The excavated material is passed through openings in the cutterhead and carried out on conveyor belts deep inside the machine and then transported out of the tunnel via the rear of the machine. Rams push the machine forward as it digs.

Gripper TBMs differ as they have two gripper plates on either side of the machine that can be controlled individually. The TBMs brace themselves against the rock on both sides and the thrust cylinders push the TBMs forward.

In addition to the likes of the large projects we have examined, machines are required to dig out small tunnels for such things as pipes, and these

DIGGER TYPES

machines are called Microtunnel Boring Machines (MTBM). As these machines are small they do not have an onboard driver but are operated remotely by people above ground.

Roadheader

These machines, also used in tunnelling projects, are equipped with a boom-mounted cutterhead covered with very sharp spiky metal blades, which spin round very fast as they cut through rock. The cutterhead can swing vertically as well as horizontally.

The machine moves through a crawler track that moves the whole machine forward into the rock. A loading device pushes the excavated rock into the machine and along a conveyor belt and out to a lorry to be removed from the site.

The first roadheader patent was applied for in 1949 and the first machines started to appear in the 1950s. They were primarily used for digging out coal in soft conditions; the cutterheads tend not to be strong enough to work through hard rock.

As these machines have the ability to work without disturbing the surrounding rock, they are much used in tunnelling and underground rather than opencast mining.

Floating diggers

It is not just on land that diggers are required; they are also needed wholly or partially underwater, in shallow seas or fresh water, for example in the deepening of channels leading to harbours and ports or simply to keep the waterways navigable.

Many of these floating diggers were originally adapted backhoe machines but now ones built for special tasks have been developed. They can be divided into two main types: the mechanical and the hydraulic dredger.

Mechanical dredgers consist of a pontoon on which is mounted the backhoe that pulls the bucket. Material is excavated using the bucket and is pulled to the dredger. The excavated material is removed, either by barges or dumped overboard. Some of these backhoes are self propelled but many are anchored.

One of the largest backhoe dredgers in operation is the Postnik Yakovlev. This vessel is from a series of three identical

DIGGER TYPES

ones; the other two are called Vitruvius and Mimar Sinan. Capable of working in seas with waves up to 2.5m in height, they measure up at just under 70 metres in length and 18 metres wide.

The bucket ladder dredger is a stationary mechanical dredger, typically with six anchor points. It is equipped with a series of buckets that are mounted on to a continuous chain, which travels through the structure of the machine. This set-up, called the ladder, works on the same principle as

an escalator.

Although they can be used for digging out a wide variety of materials, these machines have fallen out of favour in the industry owing to their high maintenance costs, noisy operating levels and the fact that they are a danger to shipping owing to their anchor lines.

The grab dredger is another form of mechanical dredger. It is usually a static piece of machinery anchored to the sea or river bed, but some machines are self-propelled. They consist of a revolving crane fitted to a barge, to which a claw grab is attached.

Various grab buckets – such as mud grabs and sand grabs – are designed for different types of material. Generally, these dredgers perform best in soft materials.

Cutter suction dredgers are hydraulic machines consisting of a pontoon upon which is a cutting device, which loosens the earth, and some form of suction to pipe the material out. These dredgers tend to work best in clays but they are adaptable enough to be used for hard rock and gravel. When it's in operation, a cutter suction dredger may be held in position by a complex system of moorings.

The two largest examples of trailing suction diggers are the Leiv Eiriksson and the Cristóbal Colón. Both of these dredgers are equipped with two suction pipes, each with a diameter of 1.3 metres, and are capable of dredging down to depths of 155 metres. The world's most powerful self-propelled cutter dredger is the Vox Maxima, at 200 metres long and 31 metres wide. It can work in depths up to 125 metres.

BELOW A suction dredger.
Credit Walter Rademacher

GEOPOTES 15

Micro To Mega

RIGHT
A Komatsu
digging on the
roadside

Micro diggers

These are the smallest and most compact diggers, able to fit through a standard house doorway and so capable of working in what might be considered inaccessible or awkward areas, including indoors.

Micro diggers can be driven on special boards through a house and out into the back garden – hence their nickname of 'doorway diggers'. They are perfect for such jobs as laying patio slabs and garden landscaping.

Weighing in at less than a ton, these machines are light enough to be towed by the average car, yet they are very powerful. With a digging depth of around 1.5 metres, micro diggers can do the work of a number of men.

Mini diggers

These machines are used for work in confined spaces where other diggers cannot go. Some of the very smallest mini diggers can work in a space less than a metre wide.

Generally the term refers to a digger between two and six tons in weight. The smaller ones have a digging depth of around 1.5 metres and a digging radius of around 2 metres, but their power output will be only around 10hp.

By contrast, the larger machines in the mini digger category will be able to dig to a depth of around 4.5 metres, will have double the digging radius of the little machines and will boast a rating of 60-70hp.

These machines are ideally suited for the smaller job where high mobility is required, such as the digging of trenches, general landscaping, garden clearance, pond digging, tree stump removal and domestic ground clearance. Due to the kind of work they are used for, it is important that mini diggers have adjustable tracks and the ability to turn within their own width.

Many of these small diggers are skid-steer, which means they steer by stopping the wheels or tracks on one side – and they are four-wheel drive, so they usually have wheels rather than

tracks. The skid steer facility acts like a brake and the digger is capable of making a tight turn or even a full, 360 degree rotation.

Full-size diggers

Full-size diggers come in after the top end of the mini diggers, with a power output of between 75hp and 220hp and a weight range from around nine to 45 tons. Due to their size, they have many more design features and options than the smaller diggers, but the basic design is the same.

In order to match work requirements, various accessories are commonly made available. A tilt bucket allows the driver to move it from side to side and so create slopes. These buckets do not have metal teeth. For penetrating rock or concrete, a heavy bucket equipped with reinforced teeth is needed, and for clearing drainage ditches a bucket shaped like a V is available. On heavy, wet soil, special clay spades are options, as are stabilisers.

Towable diggers

These machines are self-trailering, which means they can be towed on the road without a trailer and by an ordinary car as they generally weigh up to 750kg.

After first making an appearance in the 1970s, their heyday was during the 1980s and 1990s. A number of companies, such as Powerfab, Fleming, Mantis, Benford, Gopher and Roughneck, produced them.

Most of these machines have anchor legs fixed at the rear and wheels that are hydraulically raised for moving around, but with the stabilisers out the digging position becomes very wide, so towable diggers are not practical in confined spaces. This also means that they are not useful when digging out on steep slopes or clearing out ditches.

As there are so many micro and mini machines available nowadays, towable diggers have become more or less obsolete, although they do come up for sale occasionally.

Massive machines

As we saw in the previous chapter, most massive diggers are used in mines or large-scale construction projects such as the building of the Channel Tunnel, when vast amounts of soil have

to be dug out. Increasing the size of a machine has been a trend for many heavy equipment manufacturers for many years.

The largest hydraulic extractor currently working in England is the Terex O&K RH 200. First introduced in 1987, this machine became an immediate success with a choice of either diesel-driven, water-cooled engines or electric motors. This hydraulic extractor is working in the Shotton opencast mine in Northumberland.

The Scottish Coal Company has four Liebherr R 9350 mining machines in its fleet. These hydraulic excavators have a maximum reach of 16 metres and a maximum digging depth of around 10 metres, but that is tiny in comparison to some of the giants found in American and Australian mines.

All of these modern machines are built using the latest state-of-the-art technology. Quite often, part of a machine is built in a factory but the last components are added at its final destination.

Pioneers And Innovators

Some names are almost legendary in the history of earthmoving; names that are synonymous with diggers. Just as we might use the generic term Hoover when we really mean vacuum cleaner, Biro when it should be ballpoint pen or Sellotape when we're referring to sticky tape, so many people use the word JCB for a digger – so much so that JCB even appears in the Oxford English Dictionary. But JCB is, of course, a registered trademark and should only be used for machines made by that company.

JCB

JCB is now Europe's largest manufacturer of construction equipment, operating manufacturing plants on four continents and employing around 7,000 people. But the company started out from humble beginnings.

JCB stands for Joseph Cyril Bamford, the man who in 1945 founded the company by selling a farm trailer he had refurbished using a second-hand welding kit. Encouraged by that success, he moved on to designing agricultural trailers and later went on to build the first hydraulic loader, known as Mayor. This machine allowed small contractors to have access to simple powerful earthmovers, and it revolutionised the construction industry.

It was at this point that Bamford started to paint his machinery a bold yellow, following the trend started by the CAT Company. Initially, JCB machines were painted blue and red. The distinctive JCB logo first appeared in 1953 on a backhoe loader and the JCB company started fitting their typewriters with a special key to accurately portray the JCB logo. The logo slants at 18

degrees from the horizontal and 22 degrees from the vertical, because that is the angle that Bamford chose and it has remained like that ever since.

In 1957, JCB launched the Hydra-Digga, a machine that was larger than the basic digger. In total more than 2,000 of these machines were built. The Hydra-Digga design was the forerunner of the famous 3C machine, which JCB introduced in 1960.

The cab was modelled on Bamford's own Cadillac, and it came equipped with a power point for boiling a kettle and a cigarette lighter. Bamford even visited every new purchaser of a 3C and presented them with their own kettle. This publicity stunt became legendary and Bamford, realising what a success it was, began the tradition of JCB stunts.

Famously, he formed the JCB display team, which became known as the Dancing Diggers. A typical stunt involved driving a car under a digger raised up on its hydraulic arms. These displays have become more and more elaborate over the years and they really showcase the amazing abilities of the machine operators as they push their diggers to the absolute limit of their capabilities.

In the late 1960s, JCB expanded into North America, the Netherlands, Spain and Italy. In 1969, Joseph Bamford retired and Anthony Bamford, his son, took over as chairman. In the 1980s JCB built their 10,000th backhoe loader – the 100,000th machine to be built by the company – and in the following decade they launched a backhoe capable of reaching speeds of 120 mph.

Joseph Bamford was not interested in mergers or takeovers, and was alleged to have said when referring to a rival: "Why should I take them over when I can put them out of business?" His motto was: make better products more cheaply. Over the years, JCB have only taken over two firms, preferring to design their own product range and expand. The two firms they did take over were Chaseside, an agricultural equipment manufacturer, in the 1970s and, more recently, Vibromax, who specialised in compaction equipment, notably road-rollers and compactors.

Nowadays, JCB has 13 manufacturing plants on four continents as well as 650 dealers with 930 locations worldwide. More than 9,000 backhoes are produced every year: that's the

LEFT
Stump removal using a JCB 3CX

PIONEERS AND INNOVATORS

RIGHT
On a building
site. Credit Joost
J. Bakker

equivalent of one every ten minutes. In total, the company have sold more than 400,000 backhoes since they were first produced.

They come with all sorts of standard equipment, including front and rear working lights, front and rear screen wash-wipers, front and rear horn, an audible and visual warning system for the engine oil, brake system and alternator as well as a range of creature comforts. These may include a radio, a fully enclosed tinted glass cab, adjustable seats with lumbar support, seatbelt and heater. Optional extras may include air

JCB BACKHOES: FACTS AND FIGURES

- More than 14 million nuts and bolts are used to make all the backhoes in a year.
- The fastest backhoe is the JCB GT, which has a top speed of 120mph.
- All the JCB backhoes produced in one year would be able together to move 1.3 billion tons of earth.

conditioning and a heated air suspension seat. The days of working on a site in cold, uncomfortable conditions are well and truly over.

A large number of JCB products are capable of using interchangeable and replacement parts and many attachments have become available in recent years. This allows excavators to extend their capabilities away from just one task and become more versatile and economic.

Over the years, JCB have had a number of interesting contracts and orders. One of their most recent orders, gained in the face of strict competition, was for 138 JCB 4CX backhoe loaders to be delivered to the British Army in January 2011. Instead of being painted in the JCB colours, they were painted in the NATO green livery. Additional enhancements included personal weapons stowage inside the cab, convoy lighting and increased wading ability. JCB's largest ever military order was for the American military, for 800 High Mobility Engineer Excavators (HMEE). These excavators are designed for high speed and high manoeuvrability and can travel at up to 55 mph.

More recently, the company secured

LEFT
A JCB 3CII built in the 1970s. Credit Ben Coulson

an order from the Russian engineering firm MOCT to supply machines that will help build the roads, bridges and tunnels in the Sochi area on the Black Sea coast, needed for the 2014 Winter Olympics. The machines will also help construct a groundbreaking road bridge at Vladivostok in advance of the 2012 Asia Pacific Economic Co-operation summit. The order worth £23 million was a much needed boost for the workforce, who had been put on a four-day working week and had feared redundancies.

In addition to manufacturing construction equipment, JCB are also involved in the JCB Academy, a school in Rocester, Staffordshire, close to JCB's world headquarters. This education facility has as its motto 'Developing engineers and business leaders for the future.' Named after its sponsor, the school, costing £22million, opened its doors in September 2010 with its first intake of students. The aim is to take in pupils in the 14 to 19 age range from the surrounding area and it is hoped that eventually there will be a maximum of 540 students.

The school is based in the adapted and renovated grade II-listed Tutbury Mill, originally built by the industrialist Richard Arkwright in the late 18th century. In addition to the normal teaching areas there are 12 fully equipped workshops, a gym and common rooms. A business ethos surrounds all the work at the JCB Academy and students are all expected to look businesslike and wear the uniform, which includes jackets and clip-on tie or boiler suit and work boots when they are working.

CATERPILLAR AND CAT

While they share the same heritage, the Caterpillar and Cat brands are distinctly different. In the early years of Caterpillar's history, the term Cat was synonymous with Caterpillar. Both described machines and the company that built them, but in the 1950s, as the product line expanded, Cat emerged as a distinct brand in its own right.

Thirty years later, the Cat logo was introduced, becoming the main identifier for products and services and the dealers that distributed them. Today, the Cat brand is one of many owned by Caterpillar, and it represents the

largest group of products and services in earthmoving industries across the world.

The Caterpillar Company was formed in the late 19th century by Benjamin Holt and Daniel Best, who had competed against each other to build various forms of steam tractors for use in farming. In 1925 the two companies – the Holt Manufacturing Company of Stockton, California and the CL Best Gas Traction Company of San Leandro, California, merged to form the Caterpillar Tractor Company. They based themselves in Peoria, Illinois, where the Caterpillar headquarters remain.

Caterpillar did not come into the hydraulic excavator market until the 1970s. Until then, they had specialised in earthmoving and farming equipment, predominantly tractors, bulldozers and dump trucks. But in the early 1970s, a time of recession, they introduced the Cat 225 series of excavator, followed quickly by the 235s and then the 245s, which were considerably larger than the 235 series.

When Caterpillar came out of the troubled times of the early 1980s, they found that many of their former competitors had gone out of business and new companies, especially Japanese

ones, had taken their place. These new companies were very successful and for Caterpillar to survive they needed to merge with another company in order to be able to compete. They already had had a deal in the 1960s with Mitsubishi, and this was expanded.

By 1992, Caterpillar and Mitsubishi formed a joint venture, mainly for the manufacture and distribution of lift trucks and the MCFA was formed – Mitsubishi Caterpillar Forklift America. Jungheinrich joined this partnership in January 2010. While Caterpillar was expanding their forklift market, they were still involved in the manufacture of excavators. Most of their products were powered by Perkins engines, which Caterpillar purchased in late 1997 from Lucas Varity.

Unlike JCB, who were never really interested in takeovers or acquisitions, Caterpillar have increased their sales through this method. The biggest acquisition in their 85-year history was the recent deal to purchase Bucyrus. The deal, expected to close in mid-2011, will make Caterpillar the world's biggest maker of mining equipment, adding massive mining shovels and draglines to

LEFT
Caterpillar
330B Hydraulic
Excavator. Credit
M. O.Stevens

RIGHT
credit Johann H.
Addicks

its line-up of trucks and excavators.

Caterpillar are planning to preserve the Bucyrus brand. For the deal to be a success, it will depend heavily on the emerging markets of China, India and Brazil, where Caterpillar hope that continued urbanisation and growth will drive demand for minerals like coal, copper and iron ore.

KOMATSU

Komatsu is the second largest manufacturer of diggers in the world after Caterpillar, although in certain areas of the world, notably Japan, China and the Middle East, Komatsu has a larger market share than its Illinois competitor.

The name Komatsu means little pine tree but the company took its name from the Japanese city of the same name where it was founded. The Takeuchi Mining Industry established Komatsu in 1917 to manufacture mining excavators and other equipment for in-house use. Four years later, Komatsu separated from Takeuchi, became independent and set about building its own machinery.

By the early 1960s the firm had grown to the point where a new headquarters was needed, and the

Komatsu Building was constructed in Tokyo. Expansion continued and in 1963 Komatsu entered into a technological licence tie-in with Bucyrus – and so began its foray in the hydraulic excavator market. This deal was terminated some 20 years later, but during that time, Komatsu established itself as a successful manufacturer of small and medium-sized hydraulic excavators.

In 1975, the first plant for the construction of equipment outside Japan was built in Brazil and in 1990, Komatsu built a hydraulic excavator assembly plant at Osaka. Two years later, it launched a wheeled excavator on the European market.

Komatsu then moved into the production of super large size excavators, notably the PC3000, and these were built at its Rokko plant. The late 1990s and early 2000s brought a sharp decline in Komatsu's fortunes and the company fell into the red in 1999, the first time in its long history. With some restructuring it did return to profitability in the fiscal year 2000 and 2001, and to this day it continues to expand its operations all over the world. In 2008, a hybrid 360 degree excavator was launched, but

there will be more about that machine in 'The Future' chapter.

Even before the tie-in with Bucyrus, Komatsu had been technologically advanced enough to use Komatsu

LEFT
A Komatsu 210,
Credit btr

BELOW
Credit Gregory
Garnich

engines in all its models. In addition, unlike most other brands, all parts and components for their products were designed and manufactured in-house, to ensure that everything worked together as an integrated whole.

Nowadays, Komatsu has a workforce of more than 38,000 employees and more than 250 patented inventions about sophisticated hydraulics control, and the company remain international leaders in the field of excavating equipment.

LIEBHERR

The Liebherr business was established by Hans Liebherr in 1949 in the southern German town of Biberach an der Riss, Baden-Württemberg. It initially produced mobile, easy-to-assemble, and affordable tower cranes but nowadays it manufactures some of the largest mining and digging tools in the world.

Over the years, the company, which has remained a family-controlled business to this day, has expanded into a vast group of companies employing many thousands of people in locations throughout the world. The firm is currently owned by Willi and Isolde Liebherr, the son and daughter of the founder, and they manage the supervisory board, serving as chair and vice-chair respectively.

Hans Liebherr, who had been an apprentice builder as a teenager, served as an army engineer during World War II. After the war, he got together with a local blacksmith to build a prototype transportable tower crane, which he exhibited at the 1949 Frankfurt Trade Fair. Orders for the crane slowly began to arrive, and one of Germany's great post-war industrial enterprises was born. Starting with tower cranes and then expanding into aircraft parts and household appliances, Liebherr then moved into the excavator market.

He noticed that most other manufacturers were still using the cable system in their diggers or were having problems with their hydraulic cylinders. In the late 1950s, he built his first wheeled hydraulic excavator on a tricycle undercarriage and named the machine the L300. This model has a very interesting design in which the steering wheels at the front are angled and mounted very close together. The first version was in shovel form but Liebherr later designed a backhoe attachment.

In the 1960s, Liebherr launched the 900 series, which continues to be a success to this day. The initial '9' is the product identifier number – all excavators begin with '9' – and the following series of numbers represents the operating weight category for that particular excavator.

TEREX

The Euclid Road Machinery Company of Cleveland, Ohio was one of the world's leading firms manufacturing earthmoving equipment. It operated for around 30 years from the 1920s to the 1950s, after which it was bought by General Motors and became part of that group under the title Terex.

The name Terex was invented from the two Latin words 'terra', meaning earth, and 'rex': king. Terex is now the world's third largest manufacturer of construction equipment and its products are manufactured in 50 plants around the world and marketed in more than

100 countries.

Terex have expanded by moving many historic brand names over to the Terex brand. In the excavation world, O&K (Orenstein & Koppel) is one of those that has moved over and no longer operates under its own brand name. Due to this method of operation, Terex now owns more than 50 different brands. In 2000, they acquired the UK company Fermec, and moved into the compact equipment category. The models were rebranded as Terex and were painted in the company's standard off-white colour scheme.

Based in Manchester, Fermec Terex manufactured a variety of backhoes many using Perkins engines and Danfoss hydraulics. Many observers considered Fermec backhoes to be the best in the world, but it wasn't long before Terex decided to close the Manchester factory and transfer production of mini excavators to France and Italy.

JOHN DEERE

Many companies started out in the agricultural business and then moved into the construction industry, and one of these was John Deere. As a blacksmith in Vermont, USA, founder John Deere

created his company by manufacturing a steel plough in 1837. He soon gained considerable fame for his excellent workmanship and ten years after his first plough, he was producing 1,000 ploughs a year.

At that time Deere promised: "I will never put my name on a product that does not have in it the best that is in me." The agricultural side of the business continued to grow but then the company moved into the earthmoving

equipment industry.

Over the years, the John Deere logo has changed and one way to assess the age of a digger is by the design of the famous leaping deer. The first deer to appear on the logo was in fact an African deer and not the American white tail. If the deer in the logo is a four-legged version, with its front right leg kicked forward while jumping over a log with 'trademark' printed inside it, the backhoe is from 1876 to 1911. If the

made in 1968 with a more contemporary look to show a side-on silhouette with just two legs. In 2000, John Deere unveiled the latest evolution in the trademark, with the deer now tilted to look like it is leaping upward and with its front leg curved in towards its body. In addition, all angles have been sharpened.

By 2010, John Deere had moved into manufacturing diggers for the Asian market. They signed an agreement for a joint venture with Ashok Leyland, the second largest manufacturer of commercial vehicles in India and the flagship of the Hinduja Group. The two companies initially made and marketed backhoes, but the range of products manufactured through this joint venture will be expanded to include a full line of construction equipment and the products will be exported to both Ashok Leyland and John Deere markets.

Today, the John Deere construction side of the company employs more than 50,000 people and produces a range of diggers, from the 110TLB with its 43hp, four-cylinder direct injection diesel engine through to its series J machines and the largest one, the 710J with a 226hp, six-cylinder engine.

logo shows only three legs and a deer's body and includes the text 'The Trade Mark of Quality Made Famous By Good Implements', the digger is from 1911 to 1935.

To mark its centenary, the logo was altered to a shield shape and included a deer with four legs. This logo appeared from 1936 to 1949. In the 1950s, the wording on the logo was 'John Deere' on the top and 'Quality Farm Equipment' on the bottom. A further revision was

HITACHI

Hitachi are one of the leading manufacturers of hydraulic excavators in the world, with current models ranging from mini diggers at six tons, through large-wheeled ones right up to 780 ton ultra large-face shovels. Out of all the major manufacturers, they make the widest range of excavators.

Hitachi as we now know it was founded in 1910 by Namihei Odaira, who was working at the time in the electrical repair shop of a copper mine north of Tokyo. His job was to repair their machinery but he started to experiment with his own designs and brought out the first domestically produced 3.7kw electric motor. He chose the name from the two kanji characters hi, meaning 'sun' and tachi, meaning 'rise'.

In 1949, the company developed and produced the first excavator made entirely from Japanese technology. The Hitachi U05 was the first cable-operated power shovel and it was followed in 1957 by the Hitachi U106 all-purpose excavator. By 1965, the advancement in Japanese technological knowhow had progressed to such a degree that Hitachi released the UH03, a machine developed entirely through Japanese expertise.

During the 1980s, there was trade tension between Europe and Japan over the question of Japanese exports, and Hitachi resolved this problem by choosing to go into partnership with the Italian manufacturer known then as Fiat Allis. They formed the joint venture of Fiat Hitachi and manufactured their excavators at a purpose-built factory in San Mauro, near Turin. This partnership lasted until 2001, when Fiat Hitachi announced they would end the joint venture and Hitachi decided to establish a separate identity in Europe with a new

factory in the Netherlands.

Further tie-ins took place with Kubota and John Deere. Today, John Deere and Hitachi have integrated their marketing operations in North, Central and South America.

In 1987, what was then the world's largest class of super-size hydraulic excavators was released by Hitachi and ten years later this giant of a machine, the EX3500, won the prestigious Ichimura Prize. That same year the company developed the EX5500, a super-size hydraulic excavator in the 515 ton class, the largest class in the world.

Further developments have taken place, and nowadays Hitachi are producing the EX8000, one of the world's largest hydraulic excavators. This ultra large-face shovel has a 3752hp output and a maximum service weight of 780 tons.

NEW HOLLAND
New Holland are another manufacturer of diggers that have their roots in the manufacture of agricultural equipment. The company was founded in 1895 by Abe Zimmerman of New Holland, Pennsylvania.

Like John Deere before him, he started as a blacksmith, predominantly repairing and building farm machinery before moving on and selling engines, principally the German-made four-cylinder Otto. It was not long before the young man realised that he could produce something far lighter and of better quality. After building a series of prototypes, his static engine reached the stage where it was ready for sale and in 1903, the New Holland Machine Company was born.

Success followed, and Zimmerman expanded into manufacturing other forms of agricultural machinery. In 1966, New Holland produced its first backhoe, a successful machine which ensured expansion for the company. By 1986 they were a force to be reckoned with and the car manufacturer Ford, realising that they were on to a winning formula, made bids for and eventually acquired the company, renaming it Ford New Holland.

Five years later, Fiat bought New Holland from Ford and merged it with their own construction arm. By 1996, New Holland were selling around 280 different products and globally, 250 dealers were selling the company's construction machinery. Since 1999, New Holland has been a part of the brand of CNH, which is part of the Fiat Group.

In 2002, New Holland joined forces with Kobelco and they concentrated their efforts on crawler excavators trading under the name of Fiat-Kobelco. Together with the historical brand of O&K, which had been bought by Terex, a further merger was made in 2005 and New Holland Construction was formed in order to keep it separate from the agricultural side of the business.

Today, New Holland have their headquarters in Turin but their equipment is built all around the world and the company are present in more than 150 countries worldwide. The latest New Holland crawler diggers have been developed in partnership with Kobelco, many with Isuzu engines. They range from the four-cylinder E70, with an operating weight of 6850kg, up to the six-cylinder E805, with an operating weight of 82,000kg.

CASE

Case CE, also Case Construction Equipment or simply Case, is another

brand that nowadays sits together with New Holland in the CNH Global group. This group is the third largest builder of construction equipment behind Caterpillar and Komatsu but, as with New Holland, it has kept its own identity in the digger market and you will see both Case and New Holland diggers on sites.

Case Construction were founded in 1842 by Jerome Increase Case. He started out building agricultural machines, notably threshing machines, and the development of the first steam engine for agricultural use is attributed to him. At one time, the company was the world's largest producer of steam engines.

Jerome I Case died in 1891 and his brother-in-law, Stephen Bull, became president of the company, which by that time was known as JI Case Threshing Machine Co. With the advent of oil-powered engines, the Case Company realised that changes were necessary to ensure the company's continued growth.

Just before World War I, the company started to manufacture road-building machinery and during World War II, Case contributed to the war effort by making artillery shells, bombs and anti-aircraft gun carriages for the US and Allied forces. In 1957, the company launched the world's first factory-integrated backhoe loader available from a single manufacturer under a single warranty.

Case increased their standing in the excavator market in 1977 when they acquired 40 per cent of the French excavator manufacturer Poclain, and in 1996 they acquired the Austrian company of Steyr.

Case were and continue to be frontrunners in the development and improvement of equipment. In 1998, the company became the first manufacturer of backhoes to include ride control and

skid steer on their diggers. And in 2001, they unveiled their first CX excavator, with on-board intelligence features to increase operator productivity. Some of these features include soft touch controls, a sound suppression system in the cab, along with the innovative Auto Warm-up, which prevents overworking the engine before it reaches its proper operating temperature.

Like JCB, Case have realised that publicity is very important and they have a motorcycle stunt team that travel the country putting on displays at shows and construction equipment events.

BOBCAT

Just as some people refer to any form of digger as a JCB, so any kind of loader is often referred to as a Bobcat, even when it has not been manufactured by Bobcat. And in fact, Bobcat produce more than just loaders.

The company started out in America in the late 1950s, when Minnesota farmer Eddie Velo needed an agile, compact loader to clean his turkey barns. Two blacksmith brothers, Cyril and Louis Keller, worked together to meet Velo's needs and invented a small,

LEFT Bobcat 334 Excavator. Credit Norbert Schnitzler

PIONEERS AND INNOVATORS

two-wheel drive, front-end loader with a rear caster wheel, and thus the Keller Welding company was formed.

The following summer, the Melroe brothers, who owned the Melroe Manufacturing Company, invited the Kellers to show this machine at a country fair. Visitors were very impressed by it and the Melroes, seeing the potential of the machine, purchased

the rights and hired the Kellers to refine the design and put it into production. Over the following few years, various adaptations were made and eventually a four-wheel drive version called the M400 was produced.

It was on this model in 1962 that the Bobcat brand name and logo appeared for the first time, along with the colour scheme of white paint with orange

trim that remains to this day. White was chosen as it was linked to the concept of cleanliness and the word Bobcat was chosen after a marketing team had sifted through a dictionary looking for a strong and resilient-sounding animal name – lion, tiger, panther and finally bobcat. The definition in that particular dictionary described the bobcat as "tough, quick, and agile", and so the Bobcat brand was born.

In 1995, the Melroe Company were acquired by the Ingersoll-Rand Company, who then renamed the company Bobcat Company. At this time, there were 100 European dealers and Bobcat Europe came into existence. By 2002, Bobcat had produced its 50,000th mini excavator. Five years later Ingersoll-Rand put Bobcat on the market and sold it to the South Korean firm Doosan Infracore for £2.4 million.

Today, Bobcat produce diggers in five different size ranges, with the biggest weighing up to 12.5 tons and the smallest ranging between one and two tons.

VOLVO

Volvo is a name that we associate with robust cars, but the group are also one of the market leaders in excavators. In fact, they are possibly the oldest industrial company in the world, starting out in 1832.

Johan Theofron Munktell, the son of a clergyman, founded what today is Volvo Construction Equipment in Eskilstuna, Sweden. He was a great innovator and built a variety of machines, from saw frames to steam engines. In 1906, he moved into construction machinery with a steam-powered road roller and a 360 degree excavator. Only small numbers of the excavator were built.

A century after the foundation of Munktell's company, they merged with Bolinder's, an engine specialist, to form AB Bolinder-Munktell, popularly referred to as BM. After World War II, the demand for construction equipment increased and Volvo, wanting to broaden their product range and enter the area of compact equipment, acquired AB Bolinder-Munktell.

In 1991, Åkermans, a manufacturer of compact diggers, became part of the Volvo Group and in 1995, the French manufacturer Pel-Job was acquired. Pel-Job had been the first ever suppliers of compact excavators in Europe and were the market leaders. Volvo officially phased out the Pel-Job brand name five years after the acquisition.

Volvo continued their acquisitions throughout the 1990s and 2000s, and they grew rapidly in the construction equipment market. By 2001, they were ready to develop their very own backhoe, and today they have an impressive collection of excavators in their fleet: compact, wheeled and crawler, from mini through to 71 ton machines.

TAKEUCHI

Takeuchi is a brand that came out of Japan and they have been present in the UK since 1981, when the first machines were imported as Priestman Mini Mustang.

Takeuchi originated the compact tracked excavator concept as far back as 1971, and these are still the products for which they are best known. They produce more than 1,200 per month, sold all over the world.

In 1996, Takeuchi set up their UK subsidiary based in Rochdale, and in 2001 they opened their southern depot at Thatcham, Berkshire.

Takeuchi pioneered many of the features taken for granted in other makes today, such as two-speed tracking and hydraulically extending tracks. They have a wide range of diggers in their catalogue, ranging from the mini TB108, which is the smallest excavator that they produce, weighing in at 835kg, right up to the TB1140 fully hydraulic 14 ton digger. All 11 excavators that they manufacture are made in Japan.

KUBOTA

The final company that I will deal with were established in Japan in 1890, but they did not come to the UK until 1979, making Thame in Oxfordshire their headquarters.

Kubota originally manufactured cast iron water pipes in Osaka. In the 1920s, they developed and produced their first compact kerosene engines for agricultural use. In 1930s, Kubota started producing diesel engines for land use and by 1959, they had built a water-cooled vertical type diesel engine.

In conjunction with the development of their diesel engines, they developed powered equipment and their tracked mini excavators have become some of the best sellers in the UK due to their legendary reliability and unusual anti-theft security system.

With competitive markets, many manufacturers of diggers have amalgamated with others, or sometimes manufacturers of other products have gone into the digger market. Most of these manufacturers produce diggers from the micro to the standard size and are not involved in the mining or quarrying side of the business. Other notable manufacturers include Hyundai, Ammann and Yanmar, Sunward and Mustang.

Preservation

RIGHT The
only known
remaining Marion
Model 91 shovel.
Credit: Daniel
Case

Most steam shovels and early excavators have disappeared, usually cut up for scrap at the end of their working lives, although a few can still be seen in industrial museums and private collections. As with vintage and classic cars, tractors and steam trains, there is a growing interest in the preservation of our industrial heritage and its machinery.

One of the best places in the UK to see objects from our construction heritage is at the Vintage Excavator Trust Museum (VET) in Threlkeld, Cumbria. Situated some three miles from the town of Keswick, the museum is based in a disused quarry and is run by a group of paid staff and volunteers.

The granite from Threlkeld Quarry had been used for over 100 years, first to supply ballast for various railway lines then to provide stone for Manchester waterworks and, more recently, to be made into Threlkeld concrete paving stones, which were used in many northern towns. But in 1982 the quarry

ceased to be profitable and the decision was made to close it and remove everything of the slightest value.

In 1992, the Lakeland Mines and Quarries Trust negotiated a lease with the objective of developing a museum on the site. During those intervening ten years the buildings had fallen into a bad state of repair and the Trust worked hard to get the site back into some semblance of order. In 1995, they handed the site over to the Caldbeck Mining Museum, which still runs it and displays artefacts from the mining industry.

The staff would have liked to put on display a number of large objects of historical importance, but there was nowhere to house them. So, in 1998 a meeting led by Ian Hartland, now the chairman, and Philip Peacock was organised to put this situation right, and the Vintage Excavator Trust was formed.

The aims of the organisation, as stated in a document produced at that meeting, are "to acquire, preserve, restore and keep in working order, a comprehensive collection of rope-operated excavators, associated plant, machinery and equipment, literature and models and to display, demonstrate and explain to the general public, schools, industrial history societies etc their history, evolution, construction and working principles."

The VET display now stands on the extensive quarry site, where there is ample space to display the exceptional collection of vintage excavators, many of which have been donated to the VET for preservation and are awaiting restoration. Among the displays is Ian Hartland's unique collection of excavators: a Ruston-Bucyrus 38-RB face shovel, a Ruston–Bucyrus 10RB, a

Ruston–Bucyrus RB 22 drag shovel, a Ruston–Bucyrus 110RB face shovel that is being restored and reassembled prior to moving up to the main quarry, as well as a Priestman dragline.

Some people store their private collections on the site and within this collection of preserved diggers is a Ruston-Bucyrus 21-B from 1935, a Priestman Cub MK V drag shovel, a Priestman Mini Mustang – Hydraulic excavator and a Priestman Mustang Mk III – Hydraulic excavator.

The Museum of Lincolnshire Life, situated in the middle of the city of Lincoln, is another museum dedicated to preserving elements of our industrial past. It was founded in 1969 and is housed in Victorian barracks built for the North Lincoln Militia.

Over the last 40-plus years it has built up a fascinating collection of more than 250,000 items, both on display and in store. The majority of these items are based on agriculture but one machine that is still there for the time being is a Ruston Proctor Steam Navvy no 306. This machine, built in 1909, has an interesting history principally after its working life, which was spent in a

LEFT
Road rail diggers all in a row

chalkpit in Arlesey, Bedfordshire.

When the pit closed, the navvy was abandoned and lost in a pit, which became flooded with water. The area became known as a beauty spot and the pit was given the nickname 'the Blue Lagoon' as the minerals in the water gave it a deep blue hue. During the summer months, when water evaporated from the pit and the water level dropped, a rusty object would start to appear. Permission was obtained from the owners of the lake and divers went down and discovered the remains of the old steam navvy.

It was decided to rescue the object, but that proved to be a difficult and arduous task. After 12 months of negotiations with various companies and water-sport associations connected with the lake, and the unexpected discovery that the original driver of the machine was alive, well and living within a mile of the site, permission was finally granted in August 1977 for the recovery.

Once the navvy was saved, it was transferred to the Ruston-Bucyrus works, where a team of apprentices spent many hundreds of hours completely restoring it. It then passed into the care of the Museum of Lincolnshire Life but remained on display outside, unprotected from the elements and gradually being damaged again. It is hoped that in the near future, this iconic piece of machinery will be rescued once again and transferred to the VET in Threlkeld.

The world-famous North of England open air museum at Beamish, County Durham, has one example of a steam navvy: a 100 ton Ruston Bucyrus steam navvy of 1931. It originally came from near Hull but was salvaged by a group of enthusiasts and is now regarded as the largest survivor in the UK.

Keeping it at Beamish has not been without its problems. For some time there had been concerns about asbestos in the boiler lagging and the public had been prevented from entering the machine. As recently as November 2010, the whole navvy was decontaminated and it is planned for it to be made accessible to the public once the inside has been tidied up.

In addition to this one machine, Beamish also has a photographic collection of steam navvies, and the photos include a Ruston excavator

owned by Thomas Swan & Co being used during road widening between Darlington and Barton; a Ruston Bucyrus excavator at work at Shap Granite Co; and a Ruston Bucyrus excavator at Whinsike Quarry in County Durham.

Several machines have ended up in private collections, and one of the largest of these collections is based four miles or so from Redditch, Worcestershire. SE Davis & Son are in the landscaping supplies business but, as a sideline over the last 20 years, they have been collecting and preserving historic earthmoving equipment. They now have a sizeable collection that includes excavators, draglines and shovels, and they hold regular open days for the public.

Another collection in private hands is that owned by Andrew Beaulah on his farm near Beverley in Humberside. Since 1994, Beaulah has held various open days and one of the most important events in the vintage earthmoving world is the 'Andrew Beaulah's Vintage Working Day'. Demonstrations of earthmoving machinery take place and often owners

of machinery normally kept at VET in Threlkeld bring their machines to the farm, to exhibit them and demonstrate their capabilities.

The largest walking dragline excavator in western Europe has been preserved at a site near Leeds. The Bucyrus Erie BE1150 Dragline weighs in at around 1220 tons and was originally owned by British Coal. It was in service for nearly ten years at the St Aiden's open cast mine in Swillington, on the outskirts of Leeds. 'Oddball', as the machine is called due to the weird noises made by the transformers that converted it from its original American electrical system to British, was made in the USA before being moved to South Wales and ultimately Leeds in 1974.

Oddball was transferred to West Yorkshire in a fleet of lorries and it took a team of workers some 53,000 hours to reassemble it. The machine moved at a rate of 0.25mph using its two giant feet to take 2 metre-long steps. When the dragline was decommissioned, it was another 15 years before a group of volunteers calling themselves the Friends of St Aiden's decided to restore it. Oddball is now the star attraction

during their heritage open days.

In contrast, Big Muskie, the world's largest dragline ever, is no more. But the good news is that when the decision was made to cut her up for scrap, her 230 ton bucket was saved from the scrap pile to be preserved as a memorial to all the people who worked in the mines of Ohio. It is now the centrepiece of a mining display at the Miners' Memorial Park in Ohio.

The last Marion Steam Shovel known to exist in the world is the 1906-built machine now located on a road in the small American town of LeRoy, New York. This machine has become known as the LeRoy Marion. At present it stands abandoned being attacked by rust, but it is thought that it may have had an illustrious career.

The machine was bought by the General Crushed Stone Company for

use in the Le Roy quarry. This shovel was rumoured to have actually been used in the excavation of the Panama Canal, but despite lengthy investigations, this cannot be verified, although this model was definitely used on that historic job. The model plate has been removed, although the patent plate is still in place.

In February 2008, the shovel was added to the National Register of Historic Places, an American organisation that holds the official list of the nation's historic places worthy of preservation. This steam shovel is the first one to be listed in the register and, according to the information on the register, it and the surrounding five acres of land are owned by the town of LeRoy. The area around the shovel has been cleared and a chain-link fence extends across the property.

LeRoy Marion is roughly the size of a railway carriage and is carried on two large caterpillar tracks, although originally it was made with flanged wheels. The wheels were removed in either 1923 or 1924 and a kit manufactured by Marion was used to make the conversion. The working

weight of the shovel is 105 tons and the engine house measures 3 metres wide by 14 metres long with an extension on the back for coal. The shovel was taken out of operation in 1949 and driven to the road where it remains to this day.

It is not just in the UK and the USA that you can find examples of preserved machinery. One of the best places to see some huge machines is at an open air museum near the city of Dessau in Germany. Located right in the middle of a former open-pit mine, Ferropolis ('City of Iron') has five wonderful examples of huge bucket wheeled excavators, some of which worked in opencast mines in the Moscow region of Russia.

These machines have been named Medusa, Mosquito, Gemini, Big Wheel and Mad Max, and they have been adapted so that visitors can safely venture inside them.

Perhaps the oddest bit of preservation of earthmoving machinery has taken place in the mining town of Kirkenes, in the extreme north east of Norway. Here, one of the redundant buckets from an enormous excavator has been tipped over and converted into a bus shelter.

LEFT
A coal digger. Credit Calistemon

The Lighter Side

Diggers, in one form or another, have appeared for quite some time in popular culture – and JCB machines have been the subject matter of many a popular song.

Most recently, the group Nizlopi had a Christmas hit in the UK with their song *JCB* (or *JCB Song*). The single, which was released in 2005 and went straight to number one in the charts, is based on lead singer Luke Concannon's memories of a day out of school spent with his father on their slow-moving JCB, going up and down a bypass.

Seamus Moore, from Kilkenny, Ireland and now a pub landlord in Burnt Oak, Middlesex, refers to himself as the JCB Man – and with good reason. He won a talent contest at the Annual Irish Music and Dance Festival in Southport, Lancashire with his own composition, *The JCB Song*. Since then, the song has become his signature number, and when he appears on stage he is often decked out in a boiler suit.

One term that has come into common parlance is backhoe fade, or JCB fade. It's a humorous term coined by the telecommunications industry to describe the accidental cutting of a cable by a backhoe. The term comes from the sudden and initially inexplicable loss of reception experienced when a cable is accidentally dug up and damaged. So that explains the sudden disappearance of phone line, television and internet service.

Depending on the cable destroyed, service may be interrupted to just a few customers or, for a large fibre optic cable, to millions. And it can strike any time a backhoe is in operation near a cable.

In the *Teletubbies* children's TV series, one of the live action five-minute segments featuring number-counting, seen from a Teletubby's belly, involves vehicles in lines. A number of JCBs are seen in line, their hydraulics operate as if they are dancing.

There are hundreds of digger toys on the market, but one of the more

interesting ones is the Lego Technic JCB JS220 excavator. This model, which first came out in 1989, featured a working hydraulics system using pneumatics and many other features similar to the original model.

Bruder Toys make all manner of realistic looking backhoes, following the styles of the original JCB, John Deere, Caterpillar and Volvo models.

In fiction, the digger in its various guises appears in a number of books.

THE LIGHTER SIDE

Naturally, there are many young children's storybooks about diggers, but they also sometimes appear as characters. In the *Bromeliad Trilogy*, or, as it is sometimes known, *The Nome Trilogy: Truckers, Diggers and Wings*, by Terry Pratchett, there is an earthmoving machine called Jekub.

A group of tiny creatures known as Nomes live under the floorboards in a large store that is about to be reopened, and they need to defend their home. A backhoe excavator called Jekub is brought in to help them. Jekub was the Nomes's attempt at pronouncing the word JCB, a word without vowels. In America, though, a Caterpillar backhoe is used and in the Italian editions it is a Fiat Allis, so the joke on the word JCB must fall flat in those countries.

Another children's book, *Mike Mulligan and His Steam Shovel*, written by Virginia Lee Burton and first published just before World War II, features, as its title suggests, a steam shovel. Mike Mulligan is the operator, and his machine is called Mary Anne. It was obviously based on Marion steam shovels: another play on words. Together, Mike and Mary Ann dig deep canals for boats, cut mountain passes for trains and hollow out cellars for high rise buildings.

In the *Thomas the Tank Engine and Friends* TV series, a steam shovel called Ned appears. A member of the team called The Pack and working for the Sodor Construction Company, he has a big bucket and mainly carries out clearance work and quarry digging. Ned is painted brown and orange with a grey roof and a dark brown bucket, arm and chassis and yellow detailing. Originally, Ned had 'Packard & Co' written on the sides of his body but this was later changed to Sodor Construction Company.

The John Deere Company sponsored an America music and dance television show. Originally called *Bandwagon*, it became known as the *John Deere Bandwagon*.

Diggers don't just appear in music, television and literature, however; they have now moved into the leisure market. There are four theme parks called Diggerland in locations around England: Kent, Devon, Durham and Yorkshire.

At these sites, children and adults have the chance to ride in and drive different types of construction machinery including mini and giant diggers. In addition, adults can take a

90-minute JCB driving experience, using a JCB 3CX backhoe.

It's not all work for diggers. There is often a lighter side to these machines – and they frequently appear in the most unexpected and unusual places.

The Excavator Pub at Ambergate, near Belper in Derbyshire, is an interesting one. It used to have an excavator sitting on its roof but no one seems to know if the pub was named after the excavator or if the machine was put on the roof in honour of the pub's name.

In June 2010, a Finn called Jukka Mutanen – or, as he likes to be known, Excavator Mutanen – announced that he would drive across Finland in his excavator if he got 50,000 Facebook friends. As he ended up with 100,000, he couldn't renege on his promise and so began the road trip.

The distance from the north of Finland to the southern tip is just under 600 miles and would normally take somewhere around 14 to 15 hours in a car. Carrying his equipment in the excavator bucket, Mutanen drove for ten hours a day for 29 days, at the steady speed of 2.5mph.

Rallies And Shows

The construction industry, like any other, has its own set of shows, rallies, exhibitions and demonstrations, at which all manner of equipment can be admired. One such was, until recently, the trade show SED (Site Equipment and Demonstration) show, a national event that focused on plant equipment.

Having been held for over 40 years, the show was cancelled in 2010 and 2011. It found itself sandwiched between the similar Bauma and Hillhead construction trade shows and due to the economic climate, the organisers did not think it would be viable. Before its cancellation, the event had showcased the latest construction machinery, which included a good selection of diggers. It is to be hoped it will make a comeback.

In contrast, Bauma, which is held in Munich every three years, had a show in 2010 that was considered a success even though there were fewer overseas visitors due to the ash cloud from the Icelandic volcano that affected air travel that year.

One of the stars of the show was an enormous mining machine, the 4200 SM Surface Miner from the Wirtgen Group, which is normally used in opencast mining, where it cuts and crushes material. Other notable exhibitors at Bauma included Bucyrus HEX, Caterpillar, Komatsu, Liebherr-International and Takeuchi.

Hillhead is a biennial happening, with the next show due to take place at Tarmac's Hillhead Quarry, Buxton, Derbyshire in June 2012. At Hillhead 2010, a section at the north end of the site was drilled and blasted to produce a fresh rock pile for the show, which enabled large excavators to show off their capabilities.

The Construction Plant Show in Kildare is the top showcase for Ireland's construction equipment. Like Hillhead, it is a biennial show where the manufacturers go to launch their new products and show off their latest lines.

In addition to these trade shows, shows and rallies take place all over the country, especially during the summer months, where members of the public can see diggers. These shows are usually dominated by early machinery and, in particular, steam-powered vehicles are on display.

At 600 acres, probably the largest show of all is the Great Dorset

RALLIES AND SHOWS

Steam Fair, which is held just outside Blandford Forum every August bank holiday. The fair features an abundance of traction engines, tractors and steam locomotives, but among all these pieces of spectacular machinery there is always a selection of diggers.

The Future

RIGHT
Excavation works

What does the future hold for the digger industry? As with the car industry, there is increasing concern about lowering the machines' emissions and reducing fuel consumption, making engines cleaner and leaner but with the same or even greater power.

The latest piece of legislation to affect the industry is the 'Euro Stage 3A emission' document, and it follows in the footsteps of the first legislation, which came out over ten years ago and related to diesel engines both on and off road. This latest regulation is all about limiting the amount of nitrogen oxides and particulate matter that an engine can produce. Diesel engines produce relatively small amounts of carbon dioxide; however, they do emit nitrogen oxides and particulate matter.

Stage 3A has required the digger manufacturers to make some major changes in their choice of engines. For example, New Holland originally chose Cummins as a partner for their new engine design and teamed up with their fellow Fiat group company Iveco to develop new engines. By contrast, John Deere and Kubota decided to develop their own engines to meet the emissions rules, and JCB became an engine maker in its own right.

JCB now builds its own four-cylinder inline diesel engines with four valves per cylinder and a 1.1 litre per cylinder design concept; hence these engines have been classified as 444 and named the JCB 444 Dieselmax. These engines are now built at JCB's power plant in Derbyshire. Since production started in November 2004, production of the JCB444 has reached 100 units a day and the engine is now fitted to all 2CX, 3CX and 4CX JCB backhoe loaders.

In order to comply with Stage 3A regulations, vehicles have been fitted with a variety of catalytic converters and filters. There are many varieties on the market, including the catalysed diesel particulate filter or CDPF.

This is designed to purify the exhaust from diesel engines using a ceramic honeycomb-like material that reduces particulate matter, carbon monoxide and hydrocarbons. It works by simultaneously removing nitrogen oxide,

THE FUTURE

carbon monoxide and hydrocarbons from the exhaust gases. The nitrogen oxide is reduced to nitrogen and water vapour, the carbon monoxide is oxidised to form carbon dioxide and the hydrocarbons are also oxidised to create carbon dioxide and water vapour.

Stage 3B of these regulations comes into force in 2011, while Stage 4 will be introduced from 2014. It is planned that by then nitrogen oxide emissions will be reduced by 96 per cent and particulate matter by 98 per cent from 1999 levels.

All the major manufacturers are working on engines that will make them compliant up to 2015. They need to invest heavily in new technology to keep their positions in the market. For example, Komatsu have developed their ECOT3 engine range. The ECOT3 features a heavy-duty, high-pressure fuel injection system that precisely controls the amount of fuel injected into each cylinder, and a cooled exhaust gas system to return low-oxygen exhaust gases to the combustion cylinders.

Caterpillar have announced a series of investments that, over the next four years, will help to make their ACERT engines one of the best in the world

for construction equipment, and Stage 4 compliant. The theory is similar to the ECOT3 in that ACERT carefully controls the combustion process to reduce pollutant levels.

John Deere are going down a different route to become Stage 4 and Stage 5 compliant, however. They have chosen not to opt for a catalytic reduction system but to stick with engines that they had used, with modifications. Their new PowerTech diesel engines work by using a high-pressure common rail fuel system and electronic controls. PowerTech Plus engines are equipped with cooled exhaust gas recirculation and variable geometry turbochargers, along with integrated dual exhaust filters.

As well as complying with the latest legislation, the major companies are continuing to develop new machines with the latest technology. Most of the big boys have brought out, or are bringing out, diggers with laser technology that controls very accurately the digging depths.

Laser technology makes it possible to lay areas in excess of 4500 square metres in one day to an accuracy of +/-

5 mm. Many of the big machines have GPS sensors, which enable centimetre-accuracy for the positioning of the digger, and many of these are now linked to a control system that calculates the GPS position from the back of the excavator to the tip of the bucket.

In connection with this, there are a few excavator load-measuring systems in place that accurately weigh the excavated material in the bucket and display the information on a monitor in the operator's cab. Not all excavators are fitted with this technology yet, but it won't be long before it becomes standard in all new machines as the manufacturers vie to keep or improve on their places in the market.

Caterpillar have brought out the industry's first electric-drive track-type tractor. At present only bulldozers are fitted with this engine, but it is suitable for any track-type tractor, so it may come to backhoes in the near future.

The D7E has an electric-drive power train that combines a diesel engine with electric generation. It doesn't use a battery; instead it captures mechanical energy in the flywheel that is generated during braking. All the electric components are sealed to keep water out, so the unit is safe to use in damp conditions.

One of the most innovative construction machinery companies is the firm of Mecalac, the French side of the Franco-German Mecalac Ahlmann Group. They debuted a prototype diesel-electric hybrid multifunction wheeled excavator in 2009. Powered by the B3.3 engine from Cummins, it is expected that this machine will reduce fuel consumption, CO_2 emissions, and noise levels. One of the contributions to the reduction of noise levels is the stop/start function, which switches off the motor when no movement is needed.

Caterpillar have chosen their plant in Aurora, Texas for the initial production of a range of mining shovels ranging from a 125 ton model through an 800 ton machine. Production of the smaller shovel began in early 2011 and the first commercial shovels are expected to come off the production line later in the year. The larger shovels are planned to be available for purchase in 2013 through to 2014.

Caterpillar tried to enter the large digger market about ten years ago, without success, but the company now hope that sales will be stronger.

Volvo have been trying to describe what an excavator might look like in 2020, and have named the machine the Sphinx. Instead of using hydraulics to make the shovel move, Volvo have conjured up a system of fuel cells powered by electric motors for the Sphinx. The fuel cells would use hydrogen as their fuel and so would leave no exhaust, just water vapour.

Their machine would look very different to those currently available with a fully adjustable cabin for the operator. As Volvo's vision sees a fully electronic and computerised series of controls, the driver would then be able to sit anywhere and the driver's seat would slide to whichever position gives the best view needed for the task in hand.

For really dangerous tasks, the cabin would be demountable and the excavator controlled remotely from the ground. There would also be a mix of wheels and caterpillars. The norm now is for an excavator to have two caterpillar tracks but the Sphinx would have four individually controlled tracks

that adjust to any surface, making the machine extremely manoeuvrable.

To download our latest catalogue and to view
the full range of books and DVDs visit:

www.G2ent.co.uk

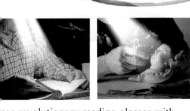
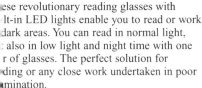

Design and Artwork by David Wildish and Scott Giarnese

Published by G2 Entertainment Limited

Publishers Jules Gammond and Edward Adams

Written by Ellie Charleston